ONE BIG BANG

Many Other BANGS

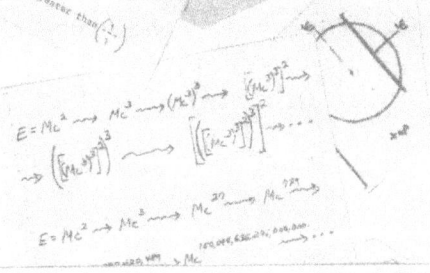

2D + 2D = 3D

The Ongoing Progression Toward 3-Dimensional Matter

By Joseph Gerchar

This book is dedicated to all of those people who have unique ideas, may they be afforded the luxury of pursuing them.

All comments are welcome

You can reach us at:

www.2plus2equals3.com

Table of contents

FORWARD

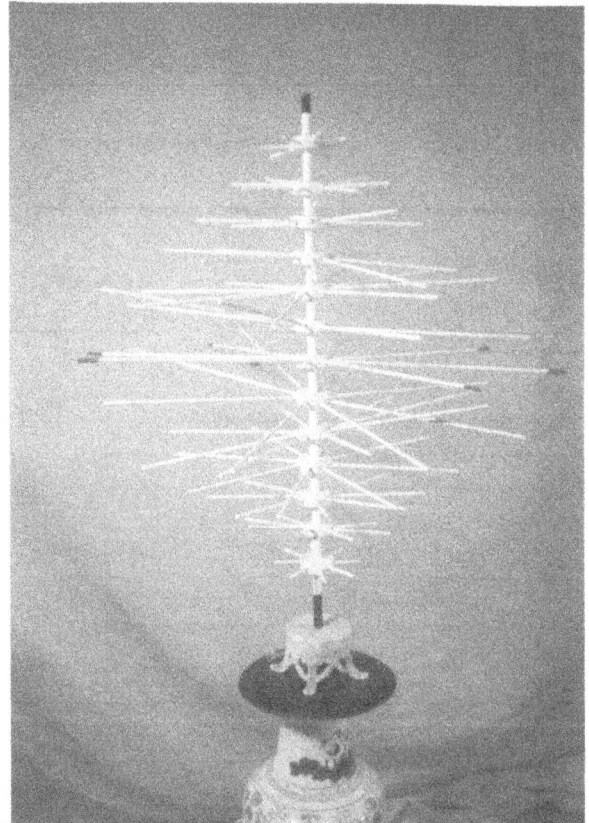

The Ongoing Progression Toward 3-Dimensional Matter

By Joseph Gerchar

I'm not a theoretical physicist, but I could theoretically be a physicist. I have thought about the expansion of the universe from time to time over the years. Not necessarily, thought about it a lot, but from time to time.

I'm not a hundred percent sure why I decided to write about it now. Perhaps it's because I just completed a three week vacation in California and I'm looking at things from a fresh perspective. Perhaps, I've been spending too much time alone in the woods. Perhaps, it's the result of having smoked too much pot in my youth that's finally catching up with me. Since I had cable in California, a rare luxury for me, perhaps, I've just been watching too many reruns of "The Big Bang Theory". I'm going to go with the latter.

On that note, when I was trying to explain this to my wife, she said that I should explain it "like Sheldon would explain it to Penny". It made me realize that she was right and I was Penny. I'm not attempting to talk down to or preach to anyone. This story is about questions, not answers. Any over-simplification is for my own benefit. It is to help me more fully understand what I'm talking about.

In an effort to avoid the over usage of the word "perhaps," or the phrase "let us consider", I, like the character Dr. Sheldon Cooper, will assume that I am always right about everything.

This book is about convincing you that some of the things that we take for granted as being correct, simply are not. I will attempt to convince you that two plus two equals three.

"The Big Bang Theory" begins as all shows begin, with the first line of the theme song. "The universe began in a hot, dense state." Well, to that let me say, "Dr. Sheldon Cooper that's just malarkey. The universe began in a cold, dense state. By the way, two plus two does not equal four. Two plus two equals three and I'm going to show you how".

INTRODUCTION

Multiple, Simultaneous,

Non-Parallel,

Proportionate Expansions

www.2plus2equals3.com

I had been thinking about Einstein. I had been thinking about E=Mc squared. I was sitting at the computer in the house and looking at a "Hubbell" telescope photo of a pulsar. I suddenly realized that I had forgotten to feed the cat.

I have a table out in the barn that I call my office and that's where the cat hangs out. I proceeded to pour milk into the plate that I kept on the floor. The milk hit the center of the plate and flowed out in all directions. Gravity began to demonstrate that my floor was not level. Because I have a big cat, I wanted to get the plate full. I slowly added more milk until gravity held one side, and the milk's own surface tension held the other. Now satisfied that the plate was full, I decided that perhaps I should work on this "Einstein" thing in my office. Being after five O-clock, I also decided that perhaps a cold beverage was in order.

I set my small cooler on the kitchen counter as my wife was removing her tea from the microwave. She set her tea down next to the cooler and proceeded to stir it. I keep a small, round plastic container in the freezer. It produces a disc of ice that is about six inches in diameter and an inch thick. As I popped the ice from the container, my wife mentioned that she didn't know where the ice pick was. Having lived on a boat for twenty-five years, I have witnessed the death of many of my favorite coolers because of Ice picks. I hadn't told her that I had hidden it in the barn the day after she bought it. I simply smiled and said, "I don't need an ice pick, I have physics."

While she watched with some annoyance, I removed the lid from the cooler. The cooler had an inch or so of cold water in it from the day before. The water was still ice cold, with a few small pieces of ice still remaining. I dropped the disc into the water. Holding up my finger I said, "Wait for it. "

After a second or two, there was a loud "pop". The disc shattered into hundreds of fragments. I lifted the disc above the surface of the water with my left hand. I removed the spoon from her tea with my other hand and lightly tapped the ice. The fragments immediately separated and fell through my fingers into the cooler. Placing the spoon back into her tea, I said "physics". Before heading back out to the barn I decided to turn the computer off. The picture of the pulsar was still on the screen.

When I returned to the barn, I could see that the cat had not been there yet. Looking down at the plate, I could see that nothing had changed. The plate, full or empty, always looked the same. Perhaps I've been looking down at the plate for so long that I simply cannot perceive it in any other way. It made me reexamine my perspective. I laid down on the floor next to the low side of the plate. I brought my eye down to where it was parallel to the plate. I could see the edge of the plate. I could see the thin line of the milk slightly above it. The milk formed a straight line, the same milk that I had poured into the center of the plate.

It made me think about the pulsar. It made me think about Einstein. Are we looking at the side of the plate? Is that what Einstein was trying to tell us by squaring the numbers?

This book is about a theory. It is a theory that is somewhat contrary to other theories. It is also a theory that I am still trying to fully understand myself.

In the course of trying to explain it, I will make mistakes. My hope is that those many people, who are more formally educated than I, will overlook them. My hope is that they will, instead, focus on the whole of the theory's content.

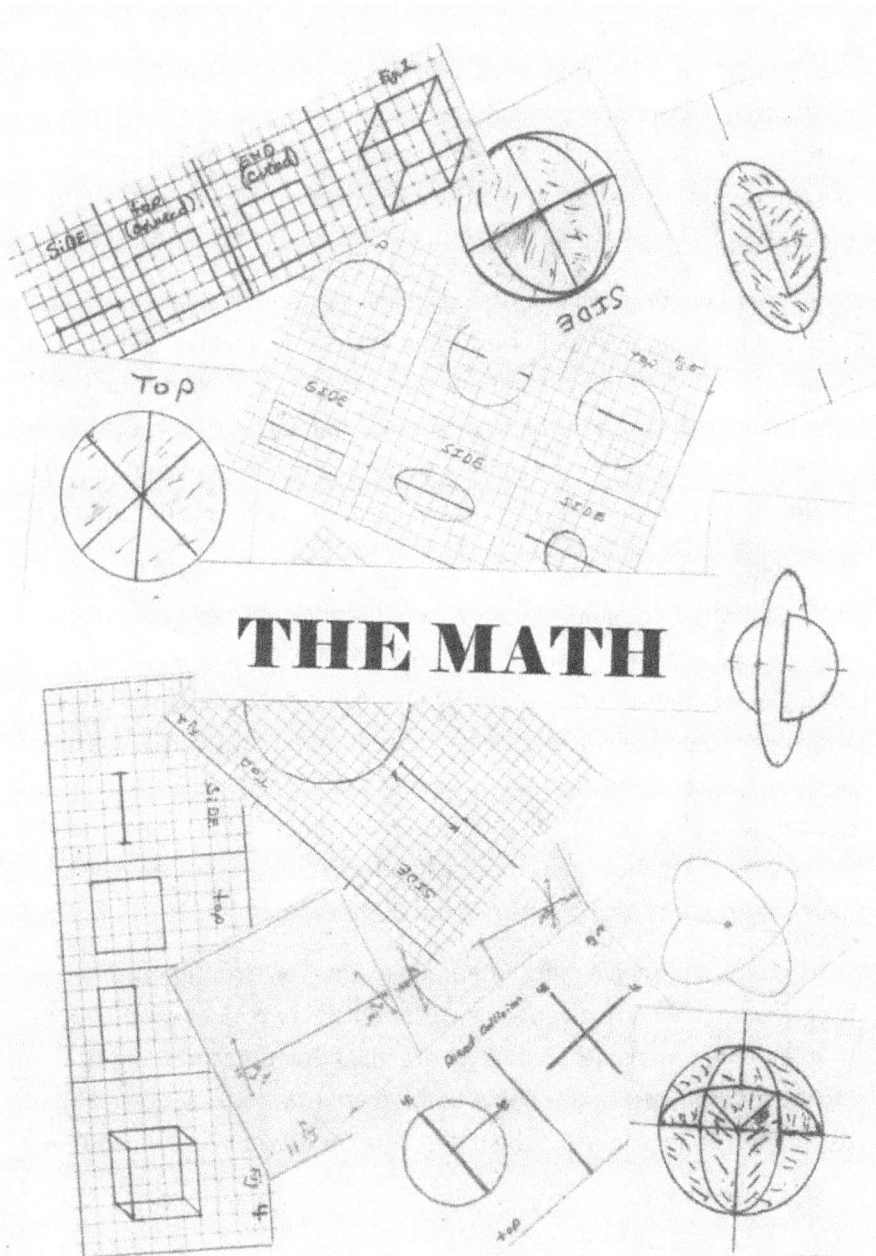

THE MATH

Like it or not, we are a species of mathematicians. Like other species that live on our planet, our brains do complicated mathematics constantly. The simple fact that we have two eyes causes our brain to perform complex calculations to determine distance, height and width. We do the math so often, and so quickly, that we are unaware of its magnitude.

When we think on a conscious level, we tend to think on a linear or two-dimensional level. For example, if we are deciding whether to go away for the weekend, we first decide to go or not. If we decide to go, then where will we go? Which car should we take? Which route do we prefer? The same process occurs if we decide not to go. What do I want to get done at home? Maybe I should wash the car. I wonder how the weather will be? Maybe I should look at the radar.

Our initial decisions trigger a two-dimensional expansion of our thought process. This process then triggers more expansions. For example, we have decided to go. We have decided where, which car, and which route we will take. Many of these decisions will now expand on to a plane of their own. We have decided to take the Mazda. Does it need new wiper blades? When was the last time that I checked the oil? We have set into motion a chain-reaction of thought processes. We have set into motion a chain-reaction of mathematics.

We also perceive mathematics in a linear or two-dimensional progression. If we are given a measurement of two inches, we know that is a linear measurement. If we are given the measurement of two inches squared, our brains know that we are now looking down at a two-dimensional square. If the measurement is two inches cubed, our brains again change perspective. As in an architectural drawing, we see a cube. (See figure 1)

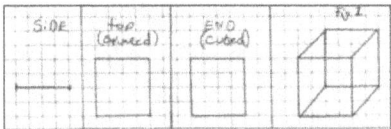

We have a little more of a problem with circles. We even had to come up with a "Transcendental Number", a non-repeating, infinite decimal to express the ratio of the circumference to the diameter of a circle.

When we take that same measurement of two inches and we say that it is one inch to the centerline, we know that we have a diameter and radius. When we square the number, we know that we are looking down at a circle or disc. If we are now told that the number is cubed our mind creates a second disc. We now see two discs that intersect, perpendicularly, at the center. We see a sphere. The added mathematical plane allows us to perceive volume. (See figure 2)

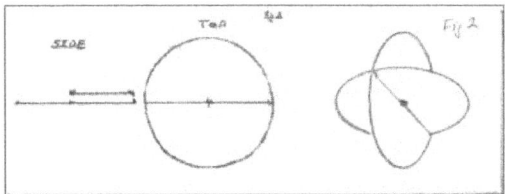

My American Heritage Dictionary defines the word "volume" as "a", the amount of space occupied by a three-dimensional object or region of space and "b", the capacity of such a region or of a specific container.

Let us take our "model" of two connecting discs that have formed our sphere, and add two more discs. The additional discs will also share the same center. They will be perpendicular to each other. They will be perpendicular to the first disc and meet the second disc at forty-five degree angles.

We still see a sphere. We still see it as having the same volume, but we also see it as more of a sphere. The addition of the two mathematical planes has improved our perception of the object. (See figure 3)

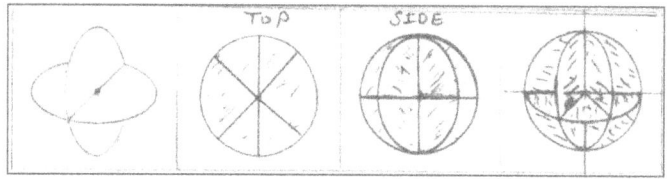

If we want to confuse our minds, let us take the second disc, and make it smaller. The original disc defines width and length. The second, now smaller disc, defines height. In the case of our square, the changing of one dimension is an easy calculation. The additional math makes the object a rectangle. (See figure 4)

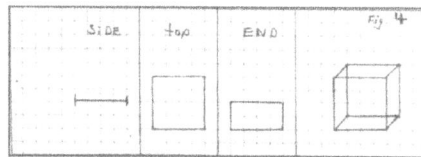

However, when we add the smaller circle, we have a problem doing the complex mathematics, necessary for us to perceive the volume of the object. (See figure 5)

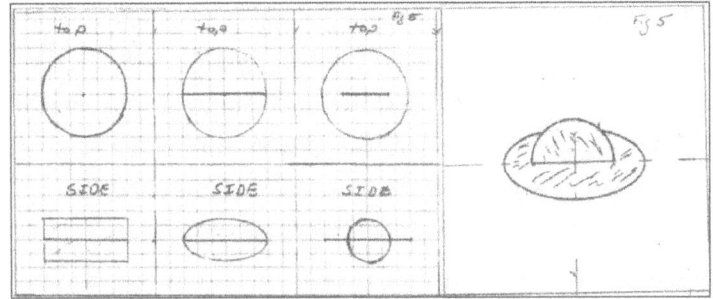

When we add the additional two, now smaller discs, we see an object that is similar to the planet Saturn. We rely on our definition of a sphere. A sphere is an object with all points equidistant from a fixed point. We simply disregard the first disc, and only see the volume of the smaller sphere. Our mind focuses on where the math is "most" complete. (See figure 6)

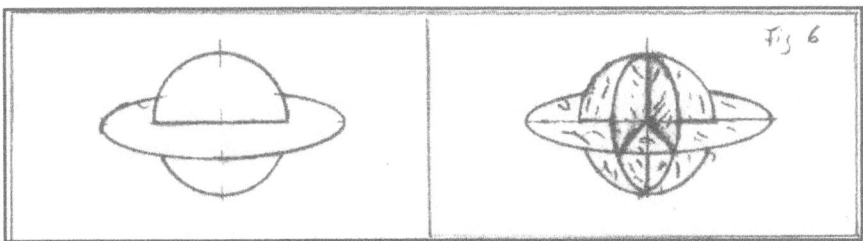

Both examples demonstrate that with additional planes of mathematics, an object is defined more clearly. They allow us to better understand the amount of space, or region of space, that the object occupies.

The additional planes also remind us of the "b" part of the dictionary's definition. They force us to see that we are not talking about volume, but rather the capacity for volume. The more discs that we add, the more we see that the volume is not yet full.

The understanding of this single, simple object, "the circle", has fueled the expansion of our understanding of mathematics. It has brought us from circle, to disc, to sphere. It has brought us from geometry, to trigonometry, to geodesic-trigonometry. It has demonstrated that the only way that we can comprehend space is to fill the space with mathematics. It has also demonstrated that it is an ongoing process.

The basis of this theory is exactly that. It is that the existence of three-dimensional matter is an ongoing process. The math is still filling the space.

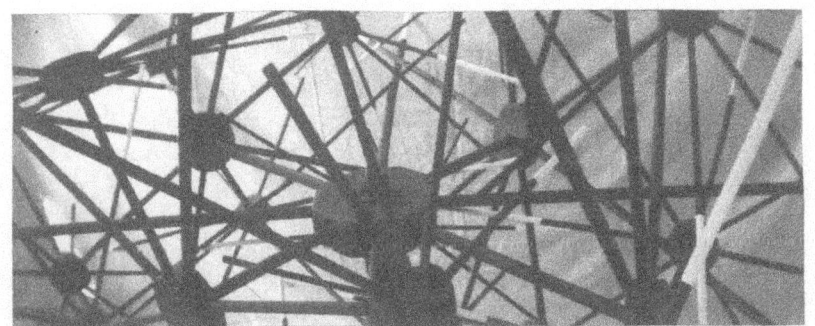

THE MODEL

This theory is about a mathematical progression. It moves from the most basic mathematical principals, to mathematics that are far beyond my ability to solve. It is a suggestion of how the mathematics might progress.

In order for me to better understand it, I have built a scale model. I call it my "Big-Boy, Tinker-Toys". "Tinker-Toys" were a collection of drilled hubs and groups of varied sized sticks. The hubs had eight holes in their sides, at forty-five degree intervals. In the center of each hub was another hole drilled perpendicular at ninety degrees from the other eight holes. By using varied sized sticks, the hubs could be connected to form three-dimensional objects. (See figure 7)

My set has 21 hubs. Unlike "Tinker-Toys", there is one large hub, 4 medium hubs and 16 smaller hubs. The different sizes were to help me to visualize the progression. In addition, unlike "Tinker-Toys", none of the center holes are drilled at ninety degrees. The holes randomly range from 85 degrees to 65 degrees. This range was determined by my ability to hand-drill holes. In reality, the angles would range from one to eighty-nine degrees. (See figure 8)

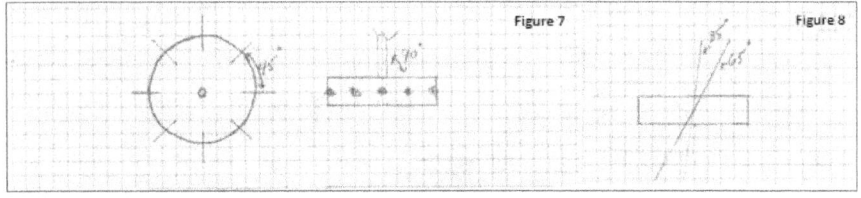

Einstein determined the scale of my model. One inch equals 186,000 miles. One inch equals the distance that light travels in one second, therefore one inch also equals one second. The model depicts the first thirty-two seconds of the expansion of the universe. It also depicts the first 11,904,000 miles of the expansion.

The model was constructed on the assumption that Einstein was correct. It also assumes that we can apply Einstein's math to the expansion of the universe. We will accept, that, when light is emitting from a central point, in opposite and parallel directions, E=Mc squared.

By accepting this, we must accept the literal description of Einstein's mathematics. We must accept the limitations and restrictions that it imposes as well.

Because E=Mc squared is referring to an ongoing progression of light moving in opposite directions, it would be more correct to say that "E" is moving toward becoming "Mc" squared. It would also be correct to say that the expansion is limited to twice the speed of light. By designating a center point, and then squaring the measurement, Einstein has given us two radii, squared. The object can only be a circle. Because it is in the process of expanding, the object can only be an expanding two-dimensional disc.

When we apply this to the "Big Bang" theory, we have a problem. If the big bang expanded outwardly in all directions into space, "E" would have to equal "Mc" cubed. Einstein's theory suggests that the big bang was a two-dimensional event. By this definition, three-dimensional matter would not exist.

My theory is based on the idea that Einstein was not just talking about a single event, but a single event that started a chain reaction of events, a chain reaction of two-dimensional events. It is based on multiple, simultaneous, non-parallel, proportionate expansions.

I have assembled my model in my barn. It is built on a base that brings the center of the first expansion about three feet off the floor. The first expansion is an octagon hub that is parallel to the ground. Of the eight holes drilled perpendicular to the center, one faces north, one south, one east, and one west. These references, along with the concept of up and down, will hopefully, help me better understand and explain the theory. (See figure 9)

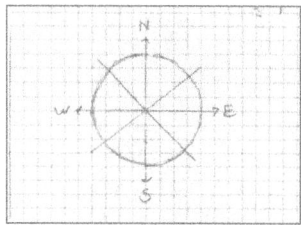

By building the model to scale, I have saved myself the trouble of doing many complex trigonometric and geodesic calculations. I can simply use a tape measure to measure one expansion in relation to another.

I had mentioned that we could confuse our minds when we give a circle height by intersecting it with a smaller, vertical circle. This model is an example of that experiment.

I believe that all research is biased. We tend to direct the progression of our research toward supporting our original premise. I believe that my ignorance of mathematics and physics, particularly of current developments in those fields, will help insulate me from being biased. I have no preconceived ideas of how this model, or the other model I built, will ultimately turn out. Both of them have been progressively assembled and have both turned out to be somewhat different than I had expected.

I will explain the assembly process that went on in the space of my model and relate that process to the space of the model's scale.

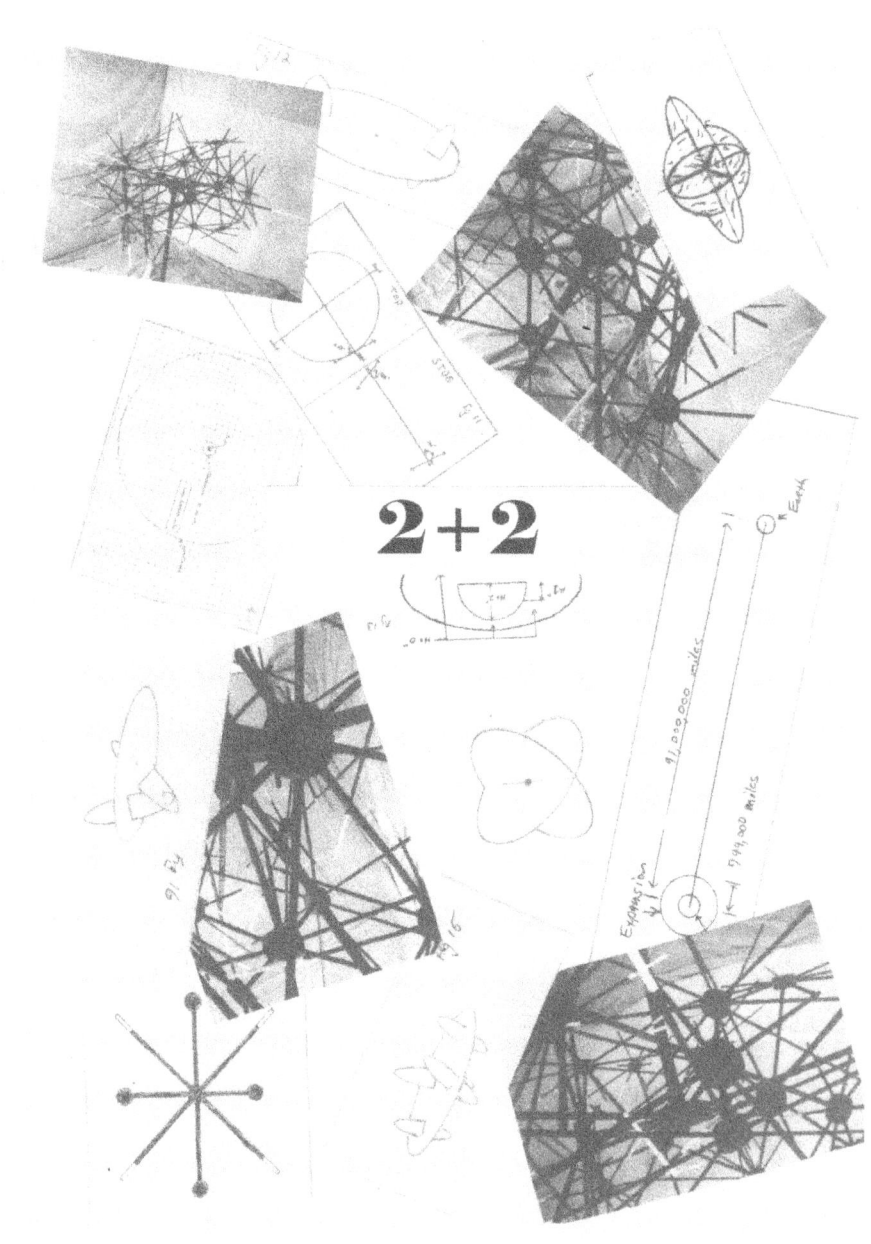

2+2

To say that it all began as a single pen point on canvas the size of our galaxy is an anthropomorphic statement. It is also a misstatement. The canvas is immensely larger than our galaxy as it's space is infinite. The single point in space is much, much smaller than a pen point. It is so small that it has only one dimension. Gravity does not yet exist. Friction does not yet exist. There is only this single one-dimensional mass, and the cold of the empty space that surrounds it.

The mass is cold beyond the zero of Celsius, that we call freezing. It is cold beyond the zero of Fahrenheit that we call intolerable. It is cold beyond the zero that we call absolute. This single, one-dimensional space contains all the matter that exists and will ever exist in the universe, and it is frozen solid.

It begins to thaw. Mathematicians might argue that it had something to do with critical mass. Theologians might say that it was the "will of God". I only know that it happened. I will not attempt to answer the question "Why". For me, the milk has already been spilled. I will attempt to answer the question "How".

"Let there be light." From our vantage point, we see two beams of light emitting in opposite and parallel directions. It has begun. We see the beams of light, as a single, straight line of infinite length. "E" is beginning its transition into becoming "MC squared".

At the designation of "squared", we must move our vantage point. We move to above, very high above. The way Einstein saw it. From our new perspective, we can see that we are looking at a disc. We see that it is expanding. We can also see, that compared to the vastness of the surrounding space, the expansion is moving very slowly.

Four seconds have passed. The diameter of the disc is now 1,488,000 miles across. To put this into perspective, if the expansion had started in the center of our sun, and the disc was on a collision course with the Earth, the disc would be now 744,000 miles closer to us. At this rate, it would be another eight minutes before it reached us. (See figure 10)

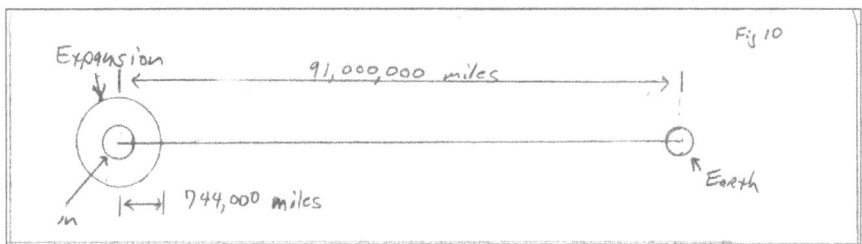

Let us, once again, change our perspective. Let us remain above, but move very close to the disc. We see that the disc is not a solid object. It is comprised of countless, smaller discs. They are all expanding. They are all thawing. They are all trying to achieve their perfect flatness. They are all trying to warm to the absolute zero that is the ambient temperature of their two-dimensional space.

The discs are crowded together. They are keeping each other cool. Their expansion is restrained. They are hindered by their own physics. They cannot exceed their own mathematics. The sum of all of them, and their plane of existence, can only expand at twice the speed of light, squared.

We also see something else; In with the discs are particles. They are fewer in number than the discs. They are being carried along with their siblings, outward at the speed of light. They are particles of one-dimensional matter. They are still frozen. The discs around them are hindering their thawing. Like a frozen turkey, left to thaw in the refrigerator, they have no concept of the ambient temperature that is outside of their space. Because of their one-dimensional state, they are tumbling. They cannot perceive the direction of the expansion of the two-dimensional plane that they are on.

Because of their lack of measurement, the particles would appear to be the same. They are not the same. They each contain different types and amounts of matter. The particles that contain less matter may only contain the matter of a few thousand galaxies. Some might contain much more. At this point in time, the amount of mass that they contain is only relevant to one thing; it is only relevant to how long it will take them to thaw.

Each of these particles has a twin. It's twin is moving in exactly the opposite direction. They are not identical twins. They may have different amounts of matter, with different densities, but they are identical in one regard. The sums of their masses, will allow both of them to thaw, at the exact same time, in a different space.

Four more seconds pass. We are now at "Big Bang", plus eight seconds. The diameter of the disc has grown to just under 3,000,000 miles. The edge is now 744,000 miles closer to the Earth.

The pair of one-dimensional particles having the least mass, have sufficiently warmed. Simultaneously, they explode. In the case of my model, there is one explosion to the north and one to the south. They both begin to expand outward. They cannot expand onto the crowded plane from which they came. There is no predetermination as to which direction they will expand. The direction will be chosen by the randomness of their tumbling through space. It is also mathematically unlikely that they will expand perpendicularly, either horizontally, or vertically, to the original plane. The math will not be that easy.

In the case of my model, the south expansion leans out on it's vertical axis at 100 degrees. The side facing west, angles toward the west at 85 degrees. The expansion to the north is angled more radically. It's vertical axis angles out at 108 degrees and it's horizontal axis angles east at 70 degrees.

In both of these expansions, the specifics of the angles are not as important as their randomness. It is important to see that the mathematical likelihood of these two discs being parallel is extremely small. (See figure 11)

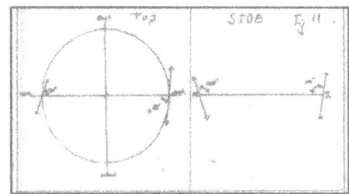

Twelve seconds after the "Big Bang "and four more seconds have passed. The original disc now has a diameter of 4,464000 miles. The Earth is still 88,768,000 miles away. The two new discs have each expanded to a diameter of 1,488,000 miles. These discs are both four seconds old. For them, time ended on the original disc, at eight seconds. Time and distance, for them, only exist on their new plane of existence.

We see that the two new discs are exactly like the original disc but smaller. Their size is relative to their youth. They are also comprised of countless smaller discs. They also each contain particles of still frozen one-dimensional matter. "E" again is moving toward becoming "Mc" squared.

On the model, we now have one primary disc that is twenty-four inches in diameter. We have two other discs, each eight inches in diameter, crossing it's radius at eight inches out from the center. We can now triangulate. The measurement of eight inches out along the radius is a constant. The measurement of four inches up from the radius is a constant. We know that the height of the newer discs diminish proportionately, as we move out to the point of intersection with the original plane. We can now calculate the relationship of any point on the radius, to any point on the smaller discs. (See figure 12)

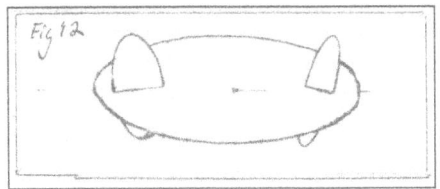

The ability to do these calculations gives us the idea that with the added measurement of height, we can now calculate volume, or at least potential volume. We must remember that we cannot. We must remember that this measurement of height only exists at the two eight inch linear lines of intersection. In all of the remaining surface area of the original disc, the height is still zero. The math has not yet filled the space. (See figure 13)

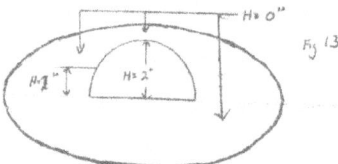

Fig 13

In addition, at this time in the model, the second pair of one-dimensional particles has thawed. They both begin to expand in the same manner. They both begin creating a linear line of intersection with the original plane.

Because my "Tinker-Toys" only have eight holes in each hub, my ability to demonstrate what is happening is limited. Of the eight sticks extending outward, from each hub, four are painted black with white ends. The other four are all black. The black sticks are divided into two pairs, with each pair being a different length. The black to white sticks represent two-dimensional light. The black sticks represent still frozen one-dimensional matter; they connect to another hub. All of the black to white sticks terminate at the limitation of the scale, at thirty-two inches long. This diminishing progression continues throughout the time frame of the scale.

In reality, the black to white sticks would fill the entire compass range, every degree, every decimal minute, every decimal second and so on. Many more holes than I could drill.

I have decided on two pairs of black sticks per progression. The mathematical progression, though at a slower speed, would happen if there was only one pair per expansion. It is my opinion, that in reality, the pairs would number in the billions. (See figure 14)

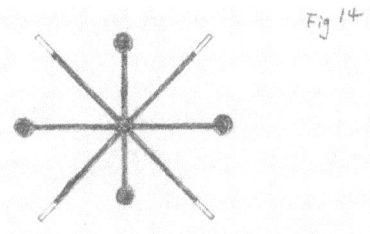

Fig 14

"Big Bang" plus sixteen seconds, the diameter of the first expansion is now 5,952,000 miles wide. The edge is now 88,024,000 miles from Earth. There are now two pairs of discs expanding on a shared axis with the original disc. Each of them is producing a linear line of intersection along that axis, at twice the speed of light. (See figure 15)

Fig 15

In the last four seconds, the light of the "Big Bang" has moved 744,000 miles closer to the Earth. At the same time, the total linear measurement of plane intersection has moved from 2,976,000 miles to 8,928000 miles.

It is also at this time that the third phase of the chain reaction begins. The two pairs of simultaneous explosions began creating four more linear lines of connection along the western and southern discs.

"Big-Bang" plus seventeen seconds, the third phase of the expansion begins on the north disc with the first pair of explosions. In the second that has passed, the original disc has increased its diameter by 372.000 miles. It is now 186,000 miles closer to the Earth. In the same second, the linear line of intersection has moved from 8,928,000 miles to 11,904,000 miles. It is a difference of 2,976,000 miles, over eight times longer than the increase in the original disc expansion, in the same amount of time.

At seventeen seconds, something else starts happening. Because of their relative angles of progression, the expanding north disc begins to collide with the expanding east disc. Their union again begins a linear expansion, in both directions, at the speed of light.

Up until now, it has been relatively easy to calculate the rate of linear growth. The math has been based only on the new plane and the plane from which it originated.

As the planes begin to have secondary connections, a second mathematical progression begins. The progression will move simultaneously with the first progression. Although they will progress in proportion to one another, the acceleration of the new progression will happen at a much faster rate. (See figure 16)

Fig 16

In the fifteen inches, or seconds remaining in my model, each of the north, south, east and west discs will produce four discs. Ultimately, including the original expansion, a total of twenty-one discs. At the end of the model the primary disc will be 11,904,000 miles wide. The disc will have progressed towards the Earth 5,952,000 miles, leaving it 85,048,000 miles away.

The linear line of primary intersections, those being from a one-dimensional matter expansion, is now 108,624,000 miles long. In addition to the primary intersections, there are sixty-three secondary intersections. It should be noted that the expansions ended after the third phase. The "Big-Bang" was the first phase. The second phase expansions occurred on average, every ten seconds. The third phase expansions occurred on average, every 9.25 seconds. By the time the model reaches thirty-two seconds, there have been no more expansions for an average of 12.75 seconds. It should also be noted that my inability to drill radically angled holes resulted in fewer, and less dramatic, secondary intersections.

A rough and very conservative measurement of the linear lines of secondary intersections is approximately 135,000,000 miles. This brings the total measurement of plane interaction to slightly over 240,000,000 miles, in thirty-two seconds. In the same amount of time, the light from the "Big-Bang" has traveled only 5,952,000 miles outward from it's point of origin. (See figure 17)

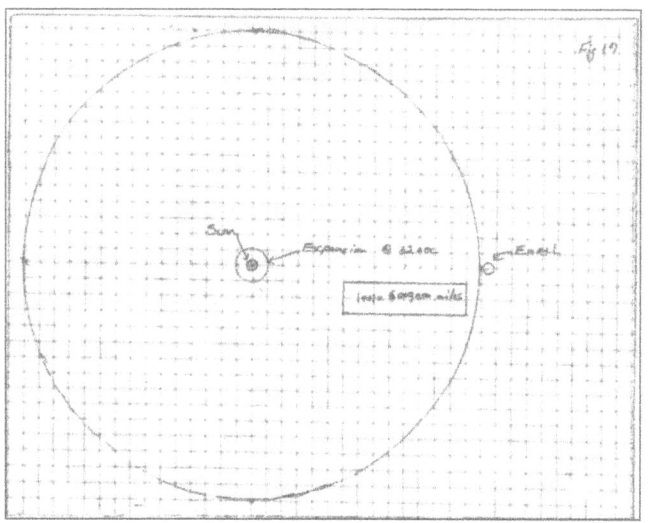

By the end of the time frame for the model, all of the twenty-one discs, except one, have made secondary intersections. In most cases, they have made many secondary intersections. If the model were to be extended by one second, to thirty-three seconds, this plane also would have begun the secondary intersection process.

If the time frame of the model were extended another three minutes, it would bring the edge of the disc to about 51,000,000 miles from Earth, not yet half way there. At this point, not only do all of the discs have secondary connections, but also each disc will intersect with every other disc. (See figure 18)

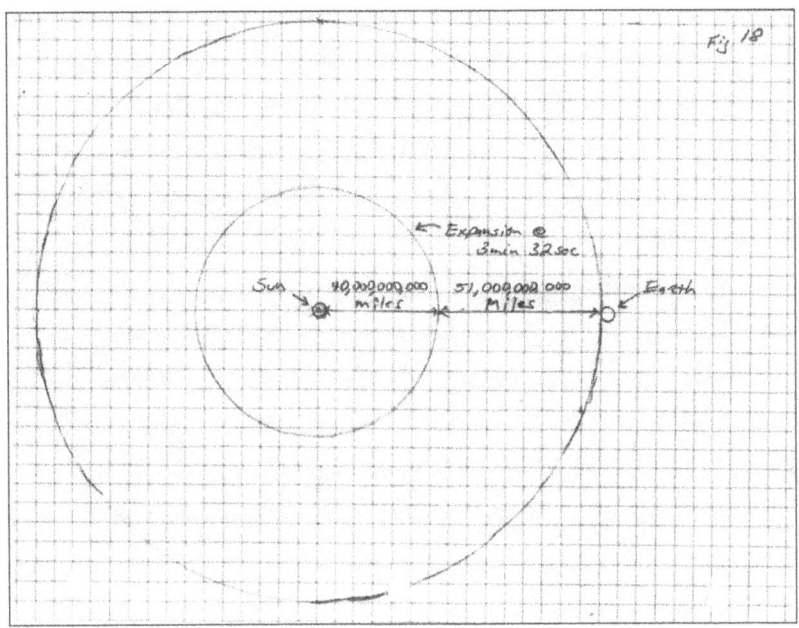

Fig 18

Sun   40,000,000,000   51,000,000,000   Earth
        miles   Miles

← Expansion @
3 min 32 sec

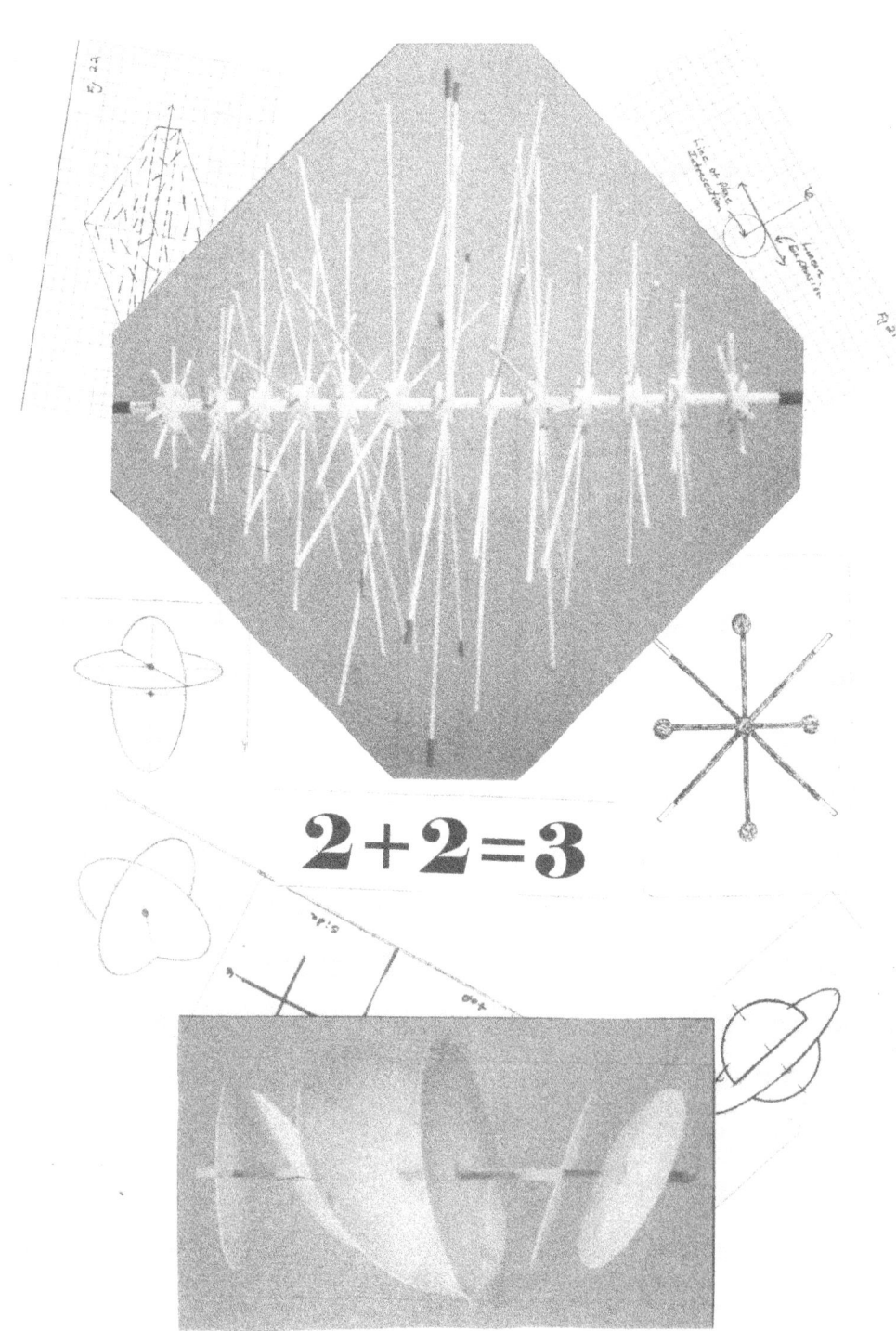

2+2=3

At the intersection of two planes, we begin a new mathematical progression. 2-D plus 2-D begins moving toward becoming 3-D. At the linear line that is created at the intersection, most of the discs from each plane, simply, miss each other. They continue outward, unaffected in their quest to achieve their perfect flatness. Any one-dimensional matter that reaches the intersection also passes through, unaffected. Its one-dimensional status makes a collision impossible.

Some of the discs collide. Those that do, fall into two categories: a direct collision or an indirect collision. A direct collision is when the discs collide in such a way that the center of their expansions meet perfectly at the linear line of intersection. At that point, they share the same center. An indirect collision refers to when the discs collide in such a way that the centers of their expansions do not share the same space on the linear line of intersection. (See Figure 19)

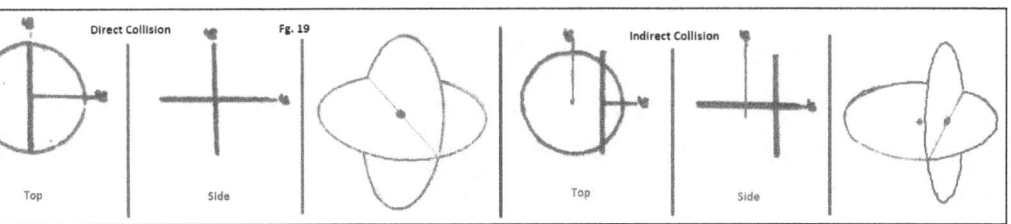

The discs that collide directly are immediately super-heated by the friction of their contact. Their temperatures soar to slightly above absolute zero. Although the two colliding discs are not the same, they are expanding at the same rate. Their proportionate growth is now locked together with the same, shared, mathematical constants. They have both learned what "up" is. For them, the math has filled the space. The once flat subatomic particles that the discs had held, all become spherical. They are no longer a part of either disc, or any plane. They are all simply set adrift into nearby space.

If we could visualize the event, we would see the line of intersection as a thin wire, a wire that was getting longer, out from its center. The direct collision of two discs would appear as a small puff of smoke emitting from the wire. As more and more collisions occur, the smoke will become thicker. At first, it would seem to be aimlessly drifting. It would then begin to acquire eddies, in different locations. The eddies are a result of the mathematical education that the particles had received when they learned what "up" is. The mathematics not only defined what "up" is, and made them spherical, it defined what volume is. It defined what gravity is. It taught them that their gravity was specific. It gave them the ability to attract, and at the same time, to hold at bay, other particles. It has allowed the nucleus to attract it's protons and it's protons to attract it's electrons. It has started another mathematical progression towards complex atoms.

The friction of their collision also warms the discs that collide indirectly. They have not warmed enough to become spherical. They are still below absolute zero. Their collision has knocked them askew. They are now too warm to squeeze back onto the disc that they came from and are too warm to enter the other disc. They are trapped between the two existences and cannot pass through. The center of both of their expansions become trapped at the line of plane intersection. They both begin to expand outward.

Although these discs contain no frozen one-dimensional matter, these discs, and all the discs from the beginning, share a common goal. They share the same mathematical inevitability. Their goal is not just to fill space on their plane, it is to fill all of the space on their plane that exists, everywhere.

I have built a second model to help me to understand this new mathematical progression. Also built of "Tinker Toys", the model has its limitations. The second model is built to the same scale as the first model, one inch equals one second, or 186,000 miles. The angles of the hubs are, again, limited to a range from 65 to 85 degrees.

The progression involves two discs, indirectly colliding. Since these collisions happen at a specific time or measurement, locating the hubs that close together was impractical. I therefore chose to portray the progression with a single disc per collision. (See Figure 20)

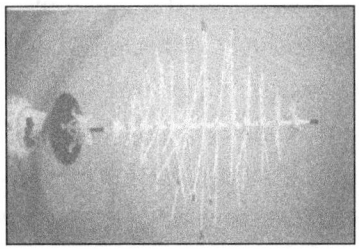

All of the hubs share the same center on a single stick, with the stick representing the linear line of contact. The intersection begins at the center, and moves outward, in both directions. The model depicts one disc being trapped, every 372,000 miles, or every two seconds.

With the model, being fourteen seconds old, the center, or oldest disc, has a fourteen-inch radius. The discs then appear outward in both directions at two-inch intervals. As they progress, each disc is, in turn, two seconds younger and 2 inches shorter than the disc that proceeded it. There are six discs progressing in each direction from the center disc, totaling 13 discs.

During the duration, or length of the model, four pairs of planes have intersected. If the model were to be extended by four seconds, or inches, all of the planes would intersect the planes that are adjacent to them. In addition, at this time, secondary intersections would begin, involving non-adjacent planes. Again, the inability to drill radically angled holes has slowed the acceleration of the progression. The model also does not depict later collisions that would occur. The spaces between the discs would begin to fill with more and younger discs.

The expanding intersection would be similar to taking a stack of paper plates that had a hole drilled through their centers, and stringing them onto an eternally growing clothesline. They all hang haphazardly, at different angles and they are all growing. The secondary intersections would first appear as a point that is away from the linear line of plane intersection. If we draw a line from that point to the linear line of intersection, the new line would expand from that point, in two directions, always perpendicular to that line. (see Figure 21)

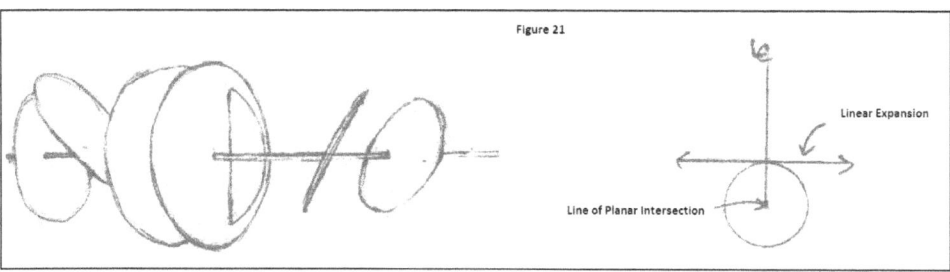
Figure 21

More and more lines would appear, expanding in different directions. The space begins to fill. At first, there would be no apparent pattern or relationship between the lines. They would just appear, always lagging behind the growth of the linear line that spawned them. As the lines get longer, and more and more appear, we began to see a structure. An object begins to take form, each line adding another piece to the geodesic puzzle.

As the math fills the space we see something that is cylindrical and at the same time conical, encompassing the linear line of intersection. Like two megaphones with their large ends connected at the site of the linear line's birth. The volume of the object fills more and more with lines, and all the lines are smoking. (See Figure 22)

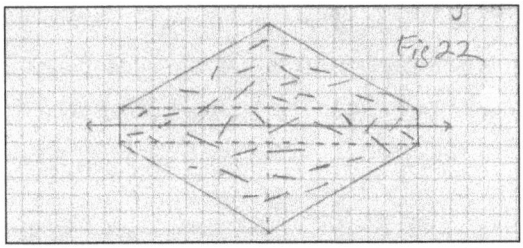

As more smoke fills the space, the movement that started as small eddies increases and becomes more and more organized. What had started as atoms gathering their particles has become the gathering of more and more atoms into more and more complex structures.

Dependent on the amount and type of raw materials available, the complex structures would vary in mass and composition. Some might be groupings of small amounts of heavy metals, such as lead or iron, with encased groupings of lighter gases. These might simply drift until the effect of a different mass increases and affects their motion. They would slowly gain a trajectory. They would start their long orbit. Their occasional collision with smaller structures would briefly thaw the frozen gasses on their surface. The gas would expand off the surface, only to be immediately refrozen and trapped by the gravitational tail of the mass.

Some groupings might be more complex. Large amounts of gasses like oxygen, hydrogen and helium might form. They might continue to draw more and more mass, until a fission reaction begins. Some of the groupings might start out as a grouping of light gasses, only to be encased by heavier and cooler atoms. These masses might appear to be cool, yet have molten centers.

As masses gather along the linear line, they draw more raw materials from it. The loose atoms spiral toward their centers, and their mass increases. It is important to remember that this process is not just happening here, at this time, in this space, but it is also happening at this time, in other spaces. It is happening at other times, in other spaces.

As different spaces become affected, their proximity becomes relevant. Masses in one space begin to affect, not only other masses in their space, but also other masses in other spaces. A new advancing mathematical progression has begun. Every new mass creates a new trajectory. More gravity becomes more motion. More motion becomes more gravity. Each new trajectory fills in another small piece of the puzzle. Each newly created orbit begins filling another space with the geodesic-trigonometry, that is it's math.

When two beams of light emit from a single point in space, in opposite but parallel directions... E=Mc^2

$$\rightarrow Mc^2$$

$$Mc^2 = E/L \rightarrow MpC^2$$

$$T/T \quad MpC^2 = 2\text{-}DMp$$

$$T/T \quad (2\text{-}DMp + 2\text{-}DMp) = 5R = H = 3\text{-}DMp$$

$$3\text{-}DMp \quad T/T \quad (3\text{-}DMp + 3\text{-}DMp) \rightsquigarrow 3\text{-}DMa$$

$$(\; 3\text{-}DMa) \rightsquigarrow 3\text{-}DMs$$

$$3\text{-}D$$

W/W 3-DMs = ($\qquad$ Ms = E/H = (3-DMa)^3

E/L $\rightsquigarrow$ (MpC^2 = 2-DM $\;$ (2-DMp + 2-DMp) $\rightarrow$

3-DMp $\rightsquigarrow$ 3-DMp + 3-DMp, $\rightsquigarrow$ 3-DMa $\rightarrow$

(3-DMa + 3-DMa) $\rightsquigarrow$ 3-DMs = E/H = (3-DMa)^3

E/L $\rightsquigarrow$ Mc^2

E/H $\rightsquigarrow$ (Mc^3)^3

$$2\text{-}D + 2\text{-}D = 3\text{-}D \qquad 3\text{-}D + 3\text{-}D = (3\text{-}D)^3$$

E/L = 2-D = Matter-times-Energy (At or below zero degrees kelvin)

E/H = 3-D = Matter-times-Energy (Above zero degrees kelvin)

E= Matter x Heat = Motion
— Moves to —

E= Matter x Heat = $\left(\dfrac{\text{Generation}}{\text{Radiation/Dissipation}}\right)$

Fig. 25

E= Matter x Heat x (Light speed 2)
= Matter x Heat x (Light speed 2)
= Matter x Heat x Motion

Line of Pulse Radionication

Fig. 26

Sub-atomic particles $\left(\dfrac{\text{Generation}}{\text{Dissipation}}\right)$ is less than $\left(\dfrac{1}{1}\right)$

$\left(\dfrac{\text{Generation}}{\text{Dissipation}}\right) = \left(\dfrac{1}{1}\right)$

Mass (Groups of Atoms) $\left(\dfrac{\text{Generation}}{\text{Dissipation}}\right)$ is greater than $\left(\dfrac{1}{1}\right)$

THE HEAT

$$E = Mc^2 \rightarrow Mc^3 \rightarrow (Mc^3)^3 \rightarrow [Mc \qquad$$

$$\rightarrow \left(\left[(Mc^3)^3\right]^3\right) \rightarrow \left[\left[(Mc^3)^3\right]^3\right]^2$$

$$E = Mc^2 \rightarrow Mc^3 \rightarrow Mc^{29} \rightarrow Mc^{29}$$

$$\rightarrow Mc^{309,429,999}$$

Fig 13
H + 0"

E/L $\rightsquigarrow$ Mc

E/H $\rightsquigarrow$ (Mc

$$2\text{-}D + 2\text{-}D = 3\text{-}$$

$$\qquad = (3\text{-}D)^3$$

E/L = 2-D = Mat er-times-Energy (At or below zero degrees kelvin)

E/H = 3-D = Matter-times-Energy (Above zero degrees kelvin)

I have a friend that had worked in the air conditioning and heating business all of his life. He told me that we cannot produce cold. He said that all we can do is move heat from one space and disperse it into another space. This results in the first space having less heat, or being colder.

All that we know how to do is make heat. Even when we want to make things cold, we do it by making heat. Our refrigerators compress a gas, which makes it hot, then releases it which makes it cool.

We have learned that we can make things hot with friction. We have moved from rubbing two sticks together, to forcing electrical energy through a resistor, creating electrical heat. We have moved from vibrating the water molecules in our food with microwaves, to colliding lead ions in the Hadron particle collider at CERN.

With that thought in mind, I decided that I needed to build a fire. The night had gotten cold, at least by Florida's standards. The temperature had plummeted into the low fifties. I live on a farm, out in the country, so I'm allowed to have a small fire. The definition of "small" is determined by my local Department of Forestry. Being civil minded, I had kept my fire within the eight foot, by eight foot, by eight foot parameters that they have set.

Since I was once a Boy Scout, I knew how to build a fire. I had decided on a tepee fire. It began with a small pile of kindling with small sticks surrounding it in the shape of a tepee. Larger and larger sticks were then added in the same fashion. My tepee was about eight feet high and eight feet wide. I carefully reached into the center and ignited the kindling. The kindling, in turn immediately ignited the pieces of wood around it. As I mentally congratulated myself, the fire grew larger. I decided to take a seat at the edge of the fire pit. After a few more moments of congratulations, I needed to move my chair back. By the time the Tepee was fully engulfed in flames, I had moved my chair back several times. I settled for a spot far enough away to be comfortable. At that spot, the radiating heat from the fire had dissipated.

The fire had provided me with some mathematical constants. If I measure the temperature at my chair, I have a starting point. If I then move toward the fire, measuring the temperature every foot as I go, I can gather enough data to calculate the rate of dissipation. If I move away from the fire, at a constant rate of speed, and take those same measurements, I can calculate the speed of the expansion. If I move away from the fire faster than the speed of its expansion, the heat ceases to exist.

Some of the mathematics on this subject have been fueled by our desire to build better bombs, in particular, nuclear bombs. We know how to calculate a blast radius. We know that we can outrun the expansion by moving away from the blast site at 20,088,000 miles per hour (3% light speed). We can calculate a diminishing mathematical progression of the dissipation of energy.

In both of these cases, we are calculating a diminishing mathematical progression. They are both diminishing at different rates, but they are both diminishing. When we think about a progression that is diminishing, we think of it like a countdown. A progression that will ultimately end in zero. Granted, a diminishing progression could be something like one divided by three, divided by three, etc., but we tend to think of a diminishing progression as finite. We can easily perceive the idea of one, divided by three, as an infinitely repeating number. We can understand the idea of "pi", an infinite non-repeating number. However, when we think about an infinite expansion, we think of an advancing mathematical progression.

I recently found a chart. It was in the form of a large thermometer that ranged from absolute zero, to absolute hot. The chart had the beginning of the universe at 1.8 decillion degrees Celsius. To write the number 1.8 decillion, you would write the number 1.8 billion, and then add twenty-four more zeros.

The chart also said that by the time the universe was one hundred seconds old, the temperature had dropped to 18 billion degrees Celsius. Because the chart gave no additional information with regard to the incremental decline in temperature, I cannot calculate the progression of energy loss. I can, however, calculate the percentage of loss and the percentage of remaining energy.

If we place the center of expansion at the center of our sun, at the speed of light, in a hundred seconds, it would have moved 18,600,000 miles toward the Earth. That's a little more than a fifth of the way there. After the same amount of time, we are left with only 180 sextillionths of one percent of the energy that we started with. One hundred and eighty sextillion would be written as 180, and 21 more zeros. It should also be pointed out that we have been dispersing this remaining 180 sextillionths of one percent, for the last 13,799,999,999 years, 364 days, 23 hours, 58 minutes, and 20 seconds. All of this is being done on a diminishing mathematical curve. I admit that my calculations might be off by a decimal point or two, but to me, it seems that we should have frozen to death long ago. (See Figure 23)

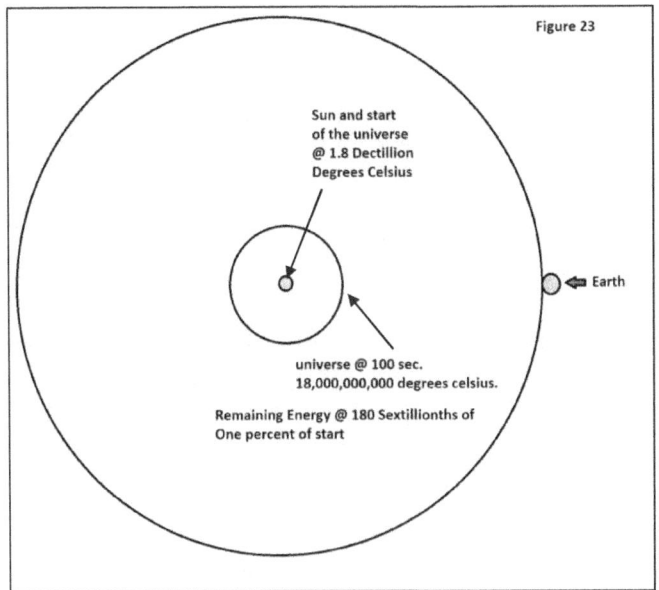

Figure 23

Sun and start
of the universe
@ 1.8 Dectillion
Degrees Celsius

Earth

universe @ 100 sec.
18,000,000,000 degrees celsius.

Remaining Energy @ 180 Sextillionths of
One percent of start

When I first started thinking about this, I realized that the most daunting part of physics or engineering equations are the acronyms. I'm guessing, that when you receive a degree in these fields, you are presented with "The Book of Acronyms". For the rest of us, footnotes would be helpful. They might actually encourage people to have more of an interest in math and science.

For the equation E=Mc squared, "E" was defined as energy, light or heat. That can be written as, E=energy, light or heat. The equation implies that energy, light and heat are different names for the same thing. However, if we rid the equation of the obvious, we could write it as: energy=light or heat. This equation could imply that energy can be either light or heat, but not both at the same time. The key word in the original definition is "or".

Let us consider writing two separate equations. In the equation E=Mc squared, squared refers to an expansion on a two-dimensional plane. For simplification, we will set that aside and focus on E=Mc. Let us write that as two separate equations, one for light and one for heat. (See Figure 24)

```
                                                        Fig. 24


    E= Matter X Light X (Light speed X 2)

    E= Matter X Heat X (Light speed X 2)

    E= Matter x Light x Motion

    E= Matter X Heat x Motion
```

Let us look at these two equations from a "human" point of view. Let's just say that we want to build a more efficient mathematics bomb. We have seen the dissipation rate for the equation, E=matter, times heat, times motion. That being said, let us start with the equation E=matter, times light, times motion.

If we use energy to create a beam of light, and shine that light into space, it will dissipate. The dissipation of energy will be at a much slower rate than if we had used the same energy to create heat. It should be noted that we are talking about white light. We have learned that if we take single frequency light, and amplify it, the dissipation rate will be drastically reduced. We call this technique "lazing" light. The invention of the laser did not create a new type of light, it was a result of better understanding an existing phenomenon.

Since we are attempting to make our bomb have an infinite blast radius, this equation seems to be the way to go. But then, there is this huge contradiction, as plain as the sunshine on our faces. In the center of our solar system is this huge star, demonstrating that E=matter, times heat, times motion.

At what point did we switch our math? More importantly, at what point did it become more efficient to switch our math? We must switch our math when E=matter, times heat, times motion, moves from being a diminishing mathematical progression to being an accelerating mathematical progression.

Let us substitute the word "motion" with the word "radiation", the word "dissipation" would also be applicable. The word "matter" is a constant in both equations, so let's set that aside. That leaves us with E=heat, times radiation. At some point in our expanding bomb the math changes. We move to E=heat, times (generation over radiation). (See Figure 25)

$$\text{Fig. 25}$$

$$E = \text{Matter} \times \text{Heat} \times \text{Motion}$$

$$-\text{Moves to}-$$

$$E = \text{Matter} \times \text{Heat} \times \left(\frac{\text{Generation}}{\text{Radiation/Dissipation}}\right)$$

This point is when three-dimensional sub atomic particles form complex atoms. At this point, the atoms themselves become generators. They and their orbiting particles become electric. Their heat is no longer dissipating. Generation over dissipation equals one over one. As more and more atoms group to form complex structures, they warm each other. The sum of their mass, produces more heat than the sum of their individual abilities to produce heat. Their joined energy moves from being electronically excited, to being vibrationally excited. Heat, times generation over dissipation, has moved from a diminishing number, to a balanced number, to a mathematically progressing number. (See Figure 26)

Fig. 26

$$\text{Sub-atomic particles} - \left(\frac{\text{Generation}}{\text{Dissipation}}\right) < \text{is less than} \left(\frac{1}{1}\right)$$

$$\text{Atoms} - \left(\frac{\text{Generation}}{\text{Dissipation}}\right) = \left(\frac{1}{1}\right)$$

$$\text{Mass (Groups of Atoms)} \left(\frac{\text{Generation}}{\text{Dissipation}}\right) > \text{is greater than} \left(\frac{1}{1}\right)$$

The acceptance of the existence of our "heat" equation presents an even deeper contradiction to our mathematics. It brings us back to E=Mc squared.

Energy equals matter, times twice the speed of light, squared. Energy equals matter, times light, times motion, squared. When two beams of light emit from the same point, in parallel, but opposite directions, E=Mc squared. It is two radii squared. It is a two-dimensional circle. It is easy for us to perceive, when we use light in our equation.

Then there is heat. We know a lot about heat. Let us again, for the moment, set aside the "squared" part. Energy equals matter, times heat, times generation over radiation. The replacement of light, with heat, seems quite reasonable. Even if the "generation over radiation" part was not progressing mathematically, as I had suggested, and was still diminishing, the equation makes sense. It is when we add the "squared" part, that there becomes a problem.

We know a lot about the way heat behaves. Again, we have a shining example in our faces every day. We know that heat dissipates in all directions. We know that in that regard, it is different than light. The heat is incapable of conforming to the behavior of light. It is mathematically impossible. The heat cannot be squared. It cannot flatten back into a two-dimensional state. It is incapable of becoming an expanding disc. The heat can only become an expanding sphere.

KEY

E/L --------------------Energy as light

E/H --------------------Energy as heat

Mp --------------------Matter as sub-atomic particles

Ma --------------------Matter as atoms

Ms --------------------Mass-groupings of atoms into more

complex structures

C ----------------------Light speed, times two

2-D --------------------Two-dimensional

3-D --------------------Three-dimensional

FR ----------------------Friction

H ----------------------Heat

HG --------------------Heat generation

RD --------------------Radiation/dissipation

$\rangle$ --------------------This is greater than $\rangle$ this

⌇⤳ ----------------Moves toward becoming

W/W --------------------When and where

T/T --------------------Then and there

EQUASION

When two beams of light emit from a single point in space, in opposite but parallel directions... $E=Mc^2$

$E \rightsquigarrow Mc^2$

$E \rightsquigarrow Mc^2 = E/L \rightsquigarrow MpC^2$

W/W $E/L = MpC^2$ T/T $MpC^2 = 2\text{-}DMp$

W/W $MpC^2 = 2\text{-}DMp$ T/T $(2\text{-}DMp + 2\text{-}DMp) = FR = H = 3\text{-}DMp$

W/W $(2\text{-}DMp + 2\text{-}DMp) = 3\text{-}DMp$ T/T $(3\text{-}DMp + 3\text{-}DMp) \rightsquigarrow 3\text{-}DMa$

$3\text{-}DMa = \left(\dfrac{HG}{RD} = \dfrac{1}{1} \right)$

W/W $3\text{-}DMa = \left(\dfrac{HG}{RD} = \dfrac{1}{1} \right)$ T/T $(3\text{-}DMa + 3\text{-}DMa) \rightsquigarrow 3\text{-}DMs$

$3\text{-}DMs = \left(\dfrac{HG}{RD} \rangle \dfrac{1}{1} \right)$

W/W $3\text{-}DMs = \left(\dfrac{HG}{RD} \rangle \dfrac{1}{1} \right)$ T/T $3\text{-}DMs = E/H = (3\text{-}DMa)^3$

$E/L \rightsquigarrow (MpC^2 = 2\text{-}DMp) \rightsquigarrow (2\text{-}DMp + 2\text{-}DMp) \rightsquigarrow$
$3\text{-}DMp \rightsquigarrow (3\text{-}DMp + 3\text{-}DMp) \rightsquigarrow 3\text{-}DMa \rightsquigarrow$
$(3\text{-}DMa + 3\text{-}DMa) \rightsquigarrow 3\text{-}DMs = E/H = (3\text{-}DMa)^3$

$E/L \rightsquigarrow Mc^2$

$E/H \rightsquigarrow (Mc^3)^3$

$2\text{-}D + 2\text{-}D = 3\text{-}D$ $3\text{-}D + 3\text{-}D = (3\text{-}D)^3$

$E/L = 2\text{-}D = $ Matter-times-Energy (At or below zero degrees kelvin)

$E/H = 3\text{-}D = $ Matter-times-Energy (Above zero degrees kelvin)

I had mentioned that I had no pre conceived ideas as to how my model and theory would ultimately turn out. They both, in fact, turned out very differently than I had expected. My intention was to try and demonstrate, to myself, how E=Mc squared could become E=Mc cubed; how a two-dimensional measurement becomes a three-dimensional measurement.

The ultimate result was that the model and the theory moved from Mc squared, to Mc cubed, to (Mc cubed) cubed. They moved from energy as light, as a diminishing mathematical progression, to energy as heat, as an accelerating mathematical progression. Again, from a "human" point of view, I believe that I have built a pretty good mathematical bomb. The expansion has moved from Mc squared, to Mc cubed, to Mc to the 27th power.

Even though we now have an infinitely accelerating mathematical progression, is that the best that we can do? Is there no other way that we can cram any more mathematics into our bomb? Does the progression of acceleration simply remain constant infinitely, or is there some way to make the math cube again?

The answer to both questions is no. No, the progression of the acceleration does not stop here, and no, we cannot cube the math again. The math must first revert back to its beginning. It must return to its original progression. Before the math can be cubed again, it must first be squared.

As two masses gather themselves in close proximity, they began to affect each other gravitationally. The smaller mass will be affected by the greater mass. A trajectory of motion will begin. One will begin moving around the other. An orbit will be established. Whether the orbit is a perfect circle or a perfect ellipse is unimportant. Whether or not the orbit is on a perfectly flat plane is irrelevant. What matters is that the orbit has a center. The orbit may not be perfectly round or perfectly flat, but the orbit is two-dimensional by nature. At the conclusion of the very first orbit, the math is complete. We can now calculate the location of the center, to any location on the circumference, or surface area of the plane. The math becomes squared. We have been provided with an additional set of mathematical constants.

As more and more objects begin orbiting the same center, they each bring with them their own set of mathematical constants. Each are on non-parallel orbits and are moving the math farther and farther toward defining spherical. The math slowly moves toward being cubed.

With more and more masses orbiting the same center, the sum of their masses becomes relevant. The sum of their gravities create a system that begins influencing other systems like them. What started with the nucleus attracting its orbiters, moves to a star attracting its orbiters which moves to a galaxy attracting its stars, and on, and on, and on... The process is always the same. It only differs with regard to its magnitude. The physics move from nanophysics, to atomic physics, to microphysics, to macrophysics, to astrophysics. The math and the physics both jump every time the phase shifts. They move every time we jump from squared to cubed. In order to calculate cubed, you must first know what squared is.

$$E = Mc^2 \longrightarrow Mc^3 \longrightarrow \left(Mc^3\right)^3 \longrightarrow \left[\left(Mc^3\right)^3\right]^2 \longrightarrow$$

$$\longrightarrow \left(\left[\left(Mc^3\right)^3\right]^2\right)^3 \longrightarrow \left[\left(\left[\left(Mc^3\right)^3\right]^2\right)^3\right]^2 \longrightarrow \ldots$$

$$E = Mc^2 \longrightarrow Mc^3 \longrightarrow Mc^{27} \longrightarrow Mc^{729} \longrightarrow$$

$$\longrightarrow Mc^{387,420,489} \longrightarrow Mc^{150,094,635,270,000,000.} \longrightarrow \ldots$$

are expanding at the same rate. Their proportionate growth
solar system, mathematical const-
heat, times motion.

ellipse is unimportan
plane is irrelevant. What
a perfect circle or a perfect
orbit is on a perfectly flat
orbit has a center. The c
flat, but the orbit is tw
of the very first or
location of the

rbit may not be perfectly
o-dimensional by nature.
it, the math is complete. W
center, to any location on t
rea of the plane. The math
rovided with an additional s

At what point did we switch our ma
point did it become more efficient to sw
our math when E=matter, times heat, tim
ishing mathematical progression, t
13,799,999 years, 364
this is h it to the same a
186,000 miles. The angles
from 65 to 85 degrees.
olves two discs, indirect
equals one
remaining 18
second
limited to a range
ogression inv

d more at
cher. The
ndividual
eing ele
erati

is the adjectiv
each to 4TeV in a part
create a temperature
heat equa
and every
specific time or m
was impractica
single disc

surface
urface

When an object is, "Hot", it
the kindling and we have fire.
ith sufficient force and speed. th
their sp

THE POETRY

mathematical progress
trajectory. More gravity becomes mor
ore gravity. Each new trajectory fills in another s
zle. Each newly created orbit begins filli
lesic-trigonometry, that is it's mat
ow to build better bombs

moves to a g
The process is always
Magnitude. The physics moves
micro physics, to macro-physics,
physics both jump every time the
from squared to cubed. I
ematics on

se together
ssion with a

ward, in both directions.
oped, every 372,000 miles, or every t

this subject has been fu
articular, nuclear bombs.
know that we can out-run
blast site at 20,088,000 mile
te a diminishing mathemat

ansion, by moving away from the word,
3% light speed). We can calcu
the dissipation of ene
aside. Th

Energy = matter, times light, times motion, squared. Th
was that the model and
ed, to (Mc cubed) cubed. Th
iminishing math

When we use light
to perceive, when we use light
its two radii squared, it is a two-
matter from the same point, in parallel, but oppo
twice the speed of light, s
emit. It is two radii squared,
uared,

Let us substitute the word, "
the word "dissipation" would also be
a constant in both equations, so let's set
section. They both eat, times radiation. At some point in our expand
We move to, E= heat, times (generat

lthough these discs contain no fro
discs, and all the discs from
hare the r

At the beginning of this book, I had mentioned that I wasn't sure why I was writing this now. After having this time to think about it, I think that it all comes down to that one sentence. "The universe began in a hot, dense state." I know that it is the first line of "The Big Bang Theory's" theme song, but I would venture to say that they weren't the first to use it, and won't be the last.

Now, I may have already convinced you that I'm not that good at math or physics, but I can assure you that I'm worse at English. Regardless, I have decided to try to break down the sentence. The universe began in a hot, dense state. In order to simplify the sentence, let us assume that it was dense. If all of the stuff that is currently in the universe is all crammed into this one space, I believe that there will be little debate that it must be pretty darn dense. That leaves us with "The universe began in a hot state." "The universe" is the subject or object of the sentence. "Began" is the verb because "The universe" is taking action. "Hot" is the adjective that modifies the noun, "state".

When an object is "hot", it contains heat. When we strike a flint, with sufficient force and speed, the friction will cause enough heat to ignite the kindling and we have a fire. When we take two lead ions and accelerate them each to 4TeV in a particle accelerator, the friction of their collision will create a temperature of 5.5 trillion degrees Celsius. We know a lot about heat, and everything that we know relates to friction. We have learned that heat equals friction.

Let us now look at the object of our sentence that begins with "The universe". It is important to note that it is not "The universes". Our object is a singular object. It is an object that is alone. It is an object which has no possibility of interaction with another object. It is an object that is incapable of friction. It is mathematically impossible for this object to be hot.

I guess that, for me, this is where my problem lies. No matter how hard I try, I keep seeing the sentence as "The universe (which couldn't possibly have been hot) began in a hot, dense state."

Whenever any new theory is presented, on any topic, it can cause hard feelings. Often, someone is offended. A new theory can cause alienation, anger, or even embarrassment. I understand that my actions could cause other actions or reactions.

At the beginning of this book, I had mentioned "The Big Bang Theory" television show. I am hoping that action will not initiate any other actions, in particular, legal actions. Let me say clearly how much I enjoy the program, and what a tremendous inspiration that it has been for me and millions of other people. The dedication and expertise of the cast and crew in producing the show have been a gift to humanity.

Since I had written some poetry in the past, I had been threatening to write a physics poem. I wanted it to be concise but yet descriptive of my theory. After weeks of deliberation as to which direction the poem should take, I realized that like the theory, there could be only one answer.

This poem is dedicated to the cast and crew of "The Big Bang Theory" for a job well done.

The universe began in a cold, dense state,

Then nearly fourteen billion years ago expansion started.

Wait!

It all began to warm.

Atoms started taking form.

Complex structures soon were born.

We built a fire.

We built an atom bomb.

Using trigonometry

to understand anomalies

that all started with the big, Bang…bang…bang…bang (Fade).

www.ingramcontent.com/pod-product-compliance
Lightning Source LLC
Chambersburg PA
CBHW070959180526
45168CB00003B/1215